CERITA MENGENAI NOMBOR

THE NUMBER STORY

SMALL BOOK ONE

ENGLISH - MALAY

---

*Numbers Teach Children
Their Number Names*

written and illustrated by

MISS ANNA

Early Reader Edition of *The Number Story 1*
Bronze Medal Winner, 2016 Wishing Shelf Book Award

Library of Congress Control Number: 2018902040

Names: Miss Anna, author.
Title: Number story : numbers teach children their number names / Miss Anna.
Description: Portland, OR: Lumpy Publishing, 2018.
Identifiers: ISBN 978-1-945977-88-6 | LCCN 2018902040
Summary: The pictures and rhymes present stories which introduce numbers 0-10.
Subjects: LCSH Numeration—English--Malay--Pictorial works--Juvenile literature. | BISAC JUVENILE NONFICTION /
Languages: English--Malay
Classification: LCC QA141.3 .M57 2018 | DDC 513—dc23

Publisher: Lumpy Publishing
Website: www.missannabooks.com
Email: missanna@missannabooks.com

Paperback: ISBN 978-1-945977-88-6
Printed in the U.S.A.   1 3 5 7 9 10 8 6 4 2

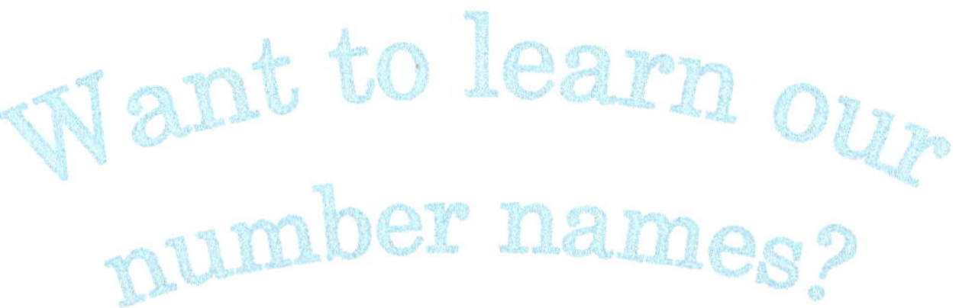

Ingin Belajar nama berkenaan Nombor?

It is very easy and a lot of fun!

Ia sangat mudah dan seronok!

Say-along our little jingle

Bernyanyilah dengan Cerpen Kami!

starting from Number One!

Ayuh, mari kita mula dengan Nombor Satu!

ONE   looks like my one finger.

SATU

seperti jari saya.

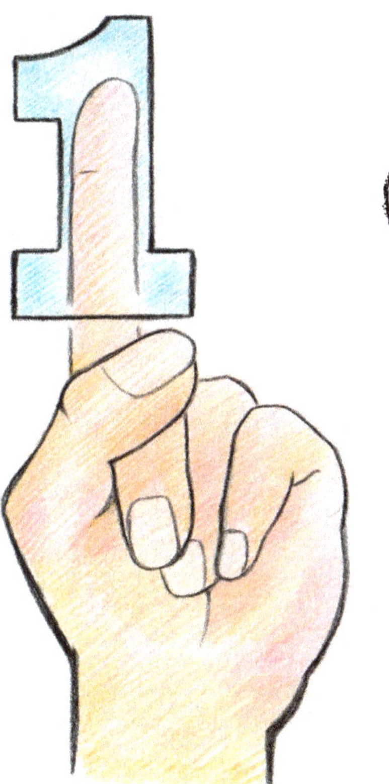
ONE!
SATU!

2

TWO   trails a tail.

DUA

pula berekor.

A TAIL! EKOR!

3

THREE has bumps.

TIGA

seperti bukit.

Ia ada lengkok.

Lihat ke lengkok ini!

Lihatlah bukit bukau yang menghijau itu!

4

FOUR carries a sail.

EMPAT

seperti kapal layar.
Ia seperti bot layar.

Sebuah bot layar!
Kapal berlayar!

5

FIVE   is a racing track.

LIMA
seperti trek perlumbaan.
Ia seperti litar lumba.

VROOM
BROOOM!
1

6
SIX curves like a snail.
ENAM
melengkung seperti siput.

A SNAIL! SIPUT!

7

SEVEN has a sharp angle.

TUJUH
seperti kapak.

Ia mempunyai satu sudut tajam.

BE CAREFUL! IT'S SHARP!
Berhati- hati! Ia Tajam!

8

EIGHT is rollercoaster rails.

LAPAN

seperti rel keretapi.

Ia seperti *roller coaster*.

YAYYY!
YIPPEE!

NINE   is a bubble on a stick.

SEMBILAN
seperti gelembung di atas tiang.

A BUBBLE!
GELEMBUNG!

TEN   is an eye of a whale.

seperti satu mata ikan Paus.

HELLO!
HELLO!

And
Dan
0
ZERO   is an empty pail.
SIFAR
seperti baldi kosong.

IT'S
EMPTY!
Ia kosong!

Thank you for playing with us today.

We had a lot of fun too!

Terima Kasih kerana bermain dengan kita hari ini.

Kami berasa seronok juga!

We are your Number friends,
Zero to Ten,
Who will be here for you~
Kami ialah kawan anda
Sifar ke Sepuluh.
Kami akan sentiasa di sini untuk anda.

Bye-bye now!
See you again soon!
Selamat tinggal!
Jumpa lagi!

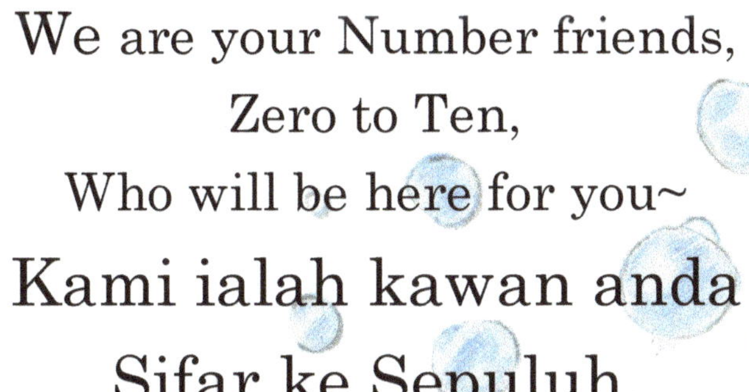

The Numbers are *SINGING* too!

To sing-a-long, look for Miss Anna Number Story
at your favorite music store like iTUNES.

MP3

Numbers 0-10
IDENTIFYING
& COUNTING

Numbers 11-20
& Ordinals
first, second, third...

Numbers 0-100
& Place Values
ones, tens, hundreds...

About Clocks
& Telling Time
hours, minutes, seconds

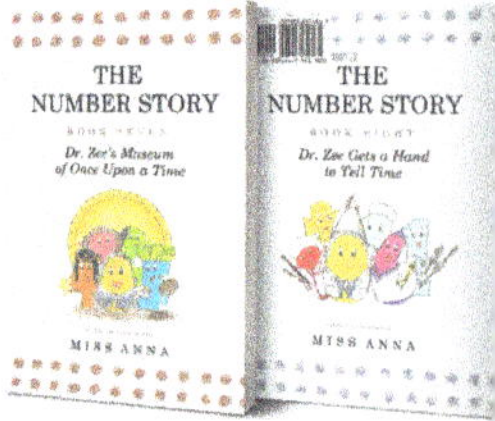

Number Story 1 & 2
isbn: 978-0-996216-48-7

Number Story 3 & 4
isbn: 978-1-945977-01-5

Number Story 5 & 6
isbn: 978-1-945977-06-0

Number Story 7 & 8
isbn: 978-1-949320-40-4

For more Miss Anna books to love,
visit us at

www.missannabooks.com

Numbers are working hard all over the world!
Come Travel the World with Us!